U.S. Department of Transportation
National Highway Traffic Safety Administration

DOT HS 811 498   July 2011

Cooperative Intersection Collision Avoidance System For Violations (CICAS-V) — Database Structure

DISCLAIMER

This publication is distributed by the U.S. Department of Transportation, National Highway Traffic Safety Administration, in the interest of information exchange. The opinions, findings, and conclusions expressed in this publication are those of the authors and not necessarily those of the Department of Transportation or the National Highway Traffic Safety Administration. The United States Government assumes no liability for its contents or use thereof. If trade names, manufacturers' names, or specific products are mentioned, it is because they are considered essential to the object of the publication and should not be construed as an endorsement. The United States Government does not endorse products or manufacturers.

REPORT DOCUMENTATION PAGE		*Form Approved* *OMB No. 0704-0188*

Public reporting burden for this collection of information is estimated to average 1 hour per response, including the time for reviewing instructions, searching existing data sources, gathering and maintaining the data needed, and completing and reviewing the collection of information. Send comments regarding this burden estimate or any other aspect of this collection of information, including suggestions for reducing this burden, to Washington Headquarters Services, Directorate for Information Operations and Reports, 1215 Jefferson Davis Highway, Suite 1204, Arlington, VA 22202-4302, and to the Office of Management and Budget, Paperwork Reduction Project (0704-0188), Washington, DC 20503.

1. AGENCY USE ONLY (Leave blank)	2. REPORT DATE July 2011	3. REPORT TYPE AND DATES COVERED October 2008 – September 2010	
4. TITLE AND SUBTITLE Cooperative Intersection Collision Avoidance System for Violations (CICAS-V) - Database Structure			5. FUNDING NUMBERS Inter-Agency Agreement
6. AUTHOR(S) Raman Sampath, Jonathan Koopmann, and Wassim G. Najm			HS-51A1 DTNH22-08-V-00017
7. PERFORMING ORGANIZATION NAME(S) AND ADDRESS(ES) U.S. Department of Transportation Research and Innovative Technology Administration John A. Volpe National Transportation Systems Center Cambridge, MA 02142			8. PERFORMING ORGANIZATION REPORT NUMBER DOT-VNTSC-NHTSA-11-07
9. SPONSORING/MONITORING AGENCY NAME(S) AND ADDRESS(ES) John Harding U.S. Department of Transportation National Highway Traffic Safety Administration Washington, DC 20590			10. SPONSORING/MONITORING AGENCY REPORT NUMBER DOT HS 811 498
11. SUPPLEMENTARY NOTES			
12a. DISTRIBUTION/AVAILABILITY STATEMENT Document is available to the public from the National Technical Information Service www.ntis.gov			12b. DISTRIBUTION CODE

13. ABSTRACT (Maximum 200 words)

This report documents the process required for data exchange between a conductor of a field operational test (FOT) and an independent evaluator based on the experience of the Cooperative Intersection Collision Avoidance System for Violations (CICAS-V) FOT project. This report also describes lessons learned from the data exchange in this project and proposes improvements to the process going forward. The main implementation of these improvements will be to design and maintain a Relational Database Management System (RDBMS). It is imperative that all future FOT conductors coordinate with the independent evaluator on the design of the RDBMS using software that will support storing, organizing, and analyzing data. The requirement should also streamline data collection, data exchange, and evaluation analyses to save resources for both the FOT conductor and the independent evaluator.

14. SUBJECT TERMS Field operational test, database, relational database management system, data acquisition system, cooperative intersection collision avoidance system for violations, data transfer, data synchronization.			15. NUMBER OF PAGES 32
			16. PRICE CODE
17. SECURITY CLASSIFICATION OF REPORT Unclassified	18. SECURITY CLASSIFICATION OF THIS PAGE Unclassified	19. SECURITY CLASSIFICATION OF ABSTRACT Unclassified	20. LIMITATION OF ABSTRACT

NSN 7540-01-280-5500

Standard Form 298 (Rev. 2-89)

Prescribed by ANSI Std. 239-18

298-102

METRIC/ENGLISH CONVERSION FACTORS

ENGLISH TO METRIC

LENGTH (APPROXIMATE)
- 1 inch (in) = 2.5 centimeters (cm)
- 1 foot (ft) = 30 centimeters (cm)
- 1 yard (yd) = 0.9 meter (m)
- 1 mile (mi) = 1.6 kilometers (km)

AREA (APPROXIMATE)
- 1 square inch (sq in, in^2) = 6.5 square centimeters (cm^2)
- 1 square foot (sq ft, ft^2) = 0.09 square meter (m^2)
- 1 square yard (sq yd, yd^2) = 0.8 square meter (m^2)
- 1 square mile (sq mi, mi^2) = 2.6 square kilometers (km^2)
- 1 acre = 0.4 hectare (he) = 4,000 square meters (m^2)

MASS - WEIGHT (APPROXIMATE)
- 1 ounce (oz) = 28 grams (gm)
- 1 pound (lb) = 0.45 kilogram (kg)
- 1 short ton = 2,000 pounds (lb) = 0.9 tonne (t)

VOLUME (APPROXIMATE)
- 1 teaspoon (tsp) = 5 milliliters (ml)
- 1 tablespoon (tbsp) = 15 milliliters (ml)
- 1 fluid ounce (fl oz) = 30 milliliters (ml)
- 1 cup (c) = 0.24 liter (l)
- 1 pint (pt) = 0.47 liter (l)
- 1 quart (qt) = 0.96 liter (l)
- 1 gallon (gal) = 3.8 liters (l)
- 1 cubic foot (cu ft, ft^3) = 0.03 cubic meter (m^3)
- 1 cubic yard (cu yd, yd^3) = 0.76 cubic meter (m^3)

TEMPERATURE (EXACT)
$[(x-32)(5/9)]$ °F = y °C

METRIC TO ENGLISH

LENGTH (APPROXIMATE)
- 1 millimeter (mm) = 0.04 inch (in)
- 1 centimeter (cm) = 0.4 inch (in)
- 1 meter (m) = 3.3 feet (ft)
- 1 meter (m) = 1.1 yards (yd)
- 1 kilometer (km) = 0.6 mile (mi)

AREA (APPROXIMATE)
- 1 square centimeter (cm^2) = 0.16 square inch (sq in, in^2)
- 1 square meter (m^2) = 1.2 square yards (sq yd, yd^2)
- 1 square kilometer (km^2) = 0.4 square mile (sq mi, mi^2)
- 10,000 square meters (m^2) = 1 hectare (ha) = 2.5 acres

MASS - WEIGHT (APPROXIMATE)
- 1 gram (gm) = 0.036 ounce (oz)
- 1 kilogram (kg) = 2.2 pounds (lb)
- 1 tonne (t) = 1,000 kilograms (kg) = 1.1 short tons

VOLUME (APPROXIMATE)
- 1 milliliter (ml) = 0.03 fluid ounce (fl oz)
- 1 liter (l) = 2.1 pints (pt)
- 1 liter (l) = 1.06 quarts (qt)
- 1 liter (l) = 0.26 gallon (gal)
- 1 cubic meter (m^3) = 36 cubic feet (cu ft, ft^3)
- 1 cubic meter (m^3) = 1.3 cubic yards (cu yd, yd^3)

TEMPERATURE (EXACT)
$[(9/5) y + 32]$ °C = x °F

QUICK INCH - CENTIMETER LENGTH CONVERSION

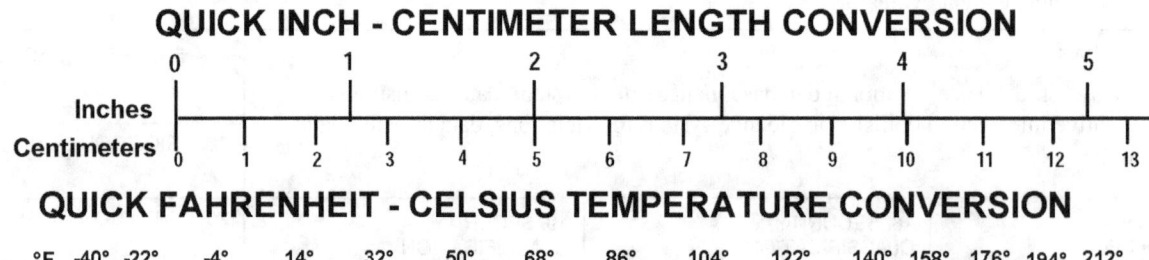

QUICK FAHRENHEIT - CELSIUS TEMPERATURE CONVERSION

°F	-40°	-22°	-4°	14°	32°	50°	68°	86°	104°	122°	140°	158°	176°	194°	212°
°C	-40°	-30°	-20°	-10°	0°	10°	20°	30°	40°	50°	60°	70°	80°	90°	100°

For more exact and or other conversion factors, see NIST Miscellaneous Publication 286, Units of Weights and Measures. Price $2.50 SD Catalog No. C13 10286

Updated 6/17/9

TABLE OF CONTENTS

EXECUTIVE SUMMARY v

1. DATA COLLECTION AND FORMAT: EXISTING PROCESS 1
    1.1. Data Acquisition System: Overview 1
    1.2. Data Collection: Data Synchronization 2
        1.2.1. Numeric Data 2
        1.2.2. Video Data 2
        1.2.3. Resolving Data Synchronization Issue 3
    1.3. Data Collection: Flat-File Format 3
    1.4. Summary: Lessons Learned 4
    1.5. Reasons to Use RDBMS 5

2. RELATIONAL DATABASE MANAGEMENT SYSTEM REQUIREMENTS FOR FOT DATABASE DESIGN AND MAINTENANCE 7
    2.1. FOT Relational Database Requirements and Design 7
    2.2. Data Exchange 10
    2.3. Data Export 10
    2.4. Data Exchange Medium 11
        2.4.1. Recommended Hardware Configuration for the SAS Interface 12

APPENDIX A. PILOT FOT DATA RECEIVED BY THE EVALUATOR 13

APPENDIX B. DATA CORRECTION PROCESS 16
    B.1. CICAS-V Data Processing Steps 16
    B.2. Data Issues 17

APPENDIX C. FINAL FOT DATABASE TABLE STRUCTURE, SCHEMA, AND IMPLEMENTATION 19
    C.1. Independent Evaluation Data Structure 19
    C.2. Data Exchange Flowchart 21

APPENDIX D. THE VOLPE CENTER'S FOT EXPERIENCE 22

LIST OF ACRONYMS

ACAS	Automotive Collision Avoidance System
ANSI	American National Standards Institute
CICAS	Cooperative Intersection Collision Avoidance
CICAS-V	Cooperative Intersection Collision Avoidance for Violations
DAS	Data Acquisition System
DDWS	Drowsy Driver Warning System
DOT	Department of Transportation
FOT	Field Operational Test
GPS	Global Positioning System
Hz	Hertz
ICC	Intelligent Cruise Control
IVBSS	Integrated-Vehicle Based Safety System
MB/s	Megabytes per second
NHTSA	National Highway Traffic Safety Administration
RDBMS	Rational Database Management System
RDCW	Roadway Departure Crash Warning System
SAS	Serial-Attached SCSI
SCSI	Small Computer System Interface
SDK	Software Development Kit
SQL	Structured Query Language
SSD	Solid State Drive
TB	Terabyte
USB	Universal Serial Bus

EXECUTIVE SUMMARY

The Cooperative Intersection Collision Avoidance System for Violations (CICAS-V) Field Operational Test (FOT) project comprises a number of teams to develop, test, and evaluate a system that warns drivers from running a red light or stop sign. The purpose of this report is to review the process required to conduct a field test involving data exchange between an FOT conductor and an independent evaluator. This report describes lessons learned from the process and proposes improvements to the process going forward. The main implementation of these improvements will be to design and maintain a Relational Database Management System (RDBMS) that will streamline efforts and resources for the independent evaluation of field test data. This report covers the following general topics:

The Process

The role of the independent evaluator is to assess safety benefits, determine driver acceptance, and quantify system performance. To conduct a successful and efficient evaluation, it is critical that data are collected by the FOT conductor and transferred to the independent evaluator in an accurate and timely manner, and that the collection method for different types of data, particularly numeric and video, results in synchronized data.

To measure driver and system performance, the FOT conductor developed a data acquisition system capable of collecting a large quantity of data from many subjects driving instrumented vehicles for an extended period of time. The FOT conductor collected primarily three kinds of data: numerical, video and audio. These data were shared with the independent evaluator.

A pilot FOT was conducted to evaluate the CICAS-V readiness for a full-scale field test and to exercise data collection and transfer processes. Due to problems with data transfer caused by data being stored in a flat-file format, further research was conducted into the development of efficient and effective relational database requirements to support data collection, data transfer, and independent evaluation.

Lessons Learned and Improvements

The system used by the FOT conductor to collect data should implement a process to synchronize the different types of data collected and data diagnostic checks for errors and duplication before transferring the data to the independent evaluator.

The flat-file format design used by the CICAS-V FOT conductor resulted in data duplication and corruption. These issues required the independent evaluator to utilize additional software and debugging routines to test for errors. After the additional steps were completed, the design of the independent evaluator's relational database was not the same as the FOT conductor's flat-file database design. This inconsistent and non-conforming file formats made data collection, testing, and analysis inefficient.

The accumulated effect of the data issues mentioned above amounts to an overall lesson learned: that the existing system and protocols are not optimized for the efficient use of time and effort by the team members.

Relational Database Management System Requirements

It is imperative that all future FOT conductors coordinate with the independent evaluator on the design of a relational database structure, using RDBMS software that will support storing, organizing, and analyzing data. The requirement should also streamline data collection, data exchange, and evaluation analyses to save resources for both the FOT conductor and the independent evaluator.

1. DATA COLLECTION AND FORMAT: EXISTING PROCESS

The success of the Cooperative Intersection Collision Avoidance System for Violations (CICAS-V) project depends on the quality and integrity of data collected by the field operational test (FOT) conductor. The Volpe National Transportation Systems Center of the U.S. Department of Transportation's Research and Innovative Technology Administration served as the independent evaluator of CICAS-V. In this role, the Volpe Center encountered many issues in the attempt to efficiently analyze the data. This section addresses the data collection process used in the CICAS-V project, the issues encountered, and a proposed relational database solution.

The subsections are as follows:

- 1.1 describes the Data Acquisition System (DAS) that was used for the pilot FOT
- 1.2 describes the issue of data collection: data synchronization
- 1.3 describes the issue of data collection: flat-file format
- 1.4 describes a summary of lessons learned
- 1.5 describes reasons to use Relational Database Management System (RDBMS)

1.1. Data Acquisition System: Overview

The FOT conductor employed a DAS[1] that was designed and developed to collect and store data. A DAS is a custom-designed, purpose-built enclosure that integrates all the signals from hardware and software components. A DAS interacts, collects, synchronizes, and stores data. The data collection process starts with periodic downloads from one or more sensors and other devices installed on instrumented vehicles and other core component hardware (e.g., single-board with embedded processor, memory, input/output add-on cards, modem, GPS, and automotive grade hard drive). The components are connected in a network and integrated to communicate with each other.

Using the DAS, data were transferred to a server at the FOT conductor's location. Data were then checked for integrity by the FOT conductor and shared among the CICAS-V teams, including the independent evaluator, over the duration of the project.

There are primarily three categories of data collected in DAS records. (Other data of lesser magnitude were also collected.):

- **Numerical data** consisting of data at various sampling rates (e.g., 10 Hz, 50 Hz) from different sensors (e.g., vehicle data bus, radar)
- **Video data** comprising frames from many video cameras at same or different sampling rates
- **Audio data** from a single source

The data collection computer runs an operating system (e.g., Windows Embedded, Linux) as well as the DAS software developed by the FOT conductor, which collects and writes data to different data files. In many cases, multiple computers are used within the DAS. For example, one computer may collect numerical data and writes it to data files while a separate computer

---

[1] A detailed insight of how the DAS is built or how it works is beyond the scope of this document.

collects video from the cameras. Synchronizing data between these two computers, which often save data and video at different rates, is of critical importance[2].

1.2. Data Collection: Data Synchronization

As stated above, the FOT conductor collected primarily numerical, video, and audio data. The data collection process must provide numeric data that are synchronized with video data; this is particularly important when the joint data are analyzed. This section reviews the data collection process for numeric and video data in detail. The key issue is that in many instances, numeric and video data were not synchronized.

1.2.1. Numeric Data

Numeric data were primarily collected and stored in many data files in the DAS by the software developed by the FOT conductor. In addition to raw data collected from the components listed previously, the DAS also computed derived values in real time. Sensor and overall system diagnostics were recorded and maintained to indicate errors such as failed sensors or communication failures. These system diagnostic and error messages are vital to help identify and understand erroneous or invalid data.

Some of the numeric data collected were time-series data. Due to time-series data being collected from most sensors at 10 Hz while radar units operated at a different update rate, an inconsistent data rate occurred. The non-synchronization of data types sometimes caused data import failures when data were exported to the independent evaluator. The different data rates sometimes caused duplicate rows of a single table (flat file) database to be populated which does not conform to relational database design standards.

Appendix A provides a detailed list of variables involved in the pilot FOT data transfer from the field test conductor to the independent evaluator.

Appendix B describes the data issues encountered in the data analysis process by the independent evaluator.

1.2.2. Video Data

Video data from many cameras were collected either at the same frequency or at different frequencies as numerical data. Video data play an important role in FOT data analysis by allowing an analyst to view the exact event conditions. Typically, video grabber hardware devices were used to collect video data. These devices support many camera sources depending on the device type. DAS developers evaluated the features, including flexibility of the device and the output before the video grabber was integrated in the DAS. Audio data were collected as data embedded in the associated video data file.

Collecting video data is one task, but when it comes to data analysis, the biggest challenge is to synchronize numerical data, such as time-series data, with the corresponding video frames. Video hardware device vendors provide a Software Development Kit (SDK) that

---

[2] Even though the DAS is built and deployed by the FOT conductor, data/video format specification and synchronization aspects must be transparent and satisfy the needs of the independent evaluator.

developers/programmers can use in their DAS software to interact with. For example, the video stream received from the device can be read and pre-processed in real time before writing to a file in a desired format.

1.2.3. Resolving Data Synchronization Issue

To resolve the data synchronization issue, the data collection process should be changed. Typically the DAS uses one or more small-footprint computers to perform specific tasks. These devices are often adjusted and configurations are changed to meet the required specifications. One computer could record numerical data and another could record the video. When a DAS is built, these computers are built on a single integrated network so that they can communicate with each other. Below are some suggestions for changing the data collection process:

- The FOT conductor may want to stamp each frame with a timestamp or with a piece of information that can be used to synchronize video with the numerical data.
- The FOT conductor may want to adjust the optimal video quality setting and collection rate that best suit the project requirements.
- Some FOT conductors may want to record individual video files. Some may record a single file that is integrated from different camera sources; for example, a quad image that integrates four cameras, but is written as a single video file.

Since independent computers record the data, synchronization is done using a piece of software written by the FOT conductor and this information is recorded as part of the data. Because this is done in real time, data errors are possible resulting in a delay in either capturing or writing the synchronization information. Alternatively, all required pieces of information may be recorded with synchronization done in the post-processing phase after the data are downloaded from test vehicles. Either way, the FOT conductor must choose a method that accurately and reliably synchronizes video with numerical data.

1.3. Data Collection: Flat-File Format

Numeric data, whether collected as raw data or derived in real time immediately after collection, were stored in a two-dimensional flat-file format, structured in rows and columns. The flat-file format used in the CICAS-V DAS was incompatible with the relational database design used by the independent evaluator.

In the CISAS-V pilot FOT, the FOT conductor and the independent evaluator did not share the same database structure. As a result, the independent evaluator had to generate several additional variables for data analysis. For this purpose, the independent evaluator created a separate relational database without altering the FOT conductor's flat-file database, which was linked to the original database for reference.

When the original data from the FOT conductor were affected with several data discrepancies, entire data had to be re-imported, which in turn affected the evaluation database and scripts. If identical databases were not maintained and there were issues with the data requiring fixes, the FOT conductor would have to resend all the affected data and the independent evaluator, in turn, would have to flush, reload, and then process the data.

Data collected in an FOT environment have a high likelihood of errors due to the uncontrolled nature of the test. Erroneous data identification and fixing methods will have to be in place to minimize the impact on a large database with millions or even billions of rows. The FOT conductor must conduct the required quality checks that will identify data discrepancies before sharing the data with the independent evaluator.

The DAS storage format and method are important because the data downloaded from the instrumented vehicles in the field test must be stored and post-processed to fix errors and to identify and eliminate unusable data. Due to the nature and size of a field test, some data loss due to DAS sensors' data, DAS software bugs, other unanticipated issues, and hardware failures are unavoidable. To limit the data loss, the FOT conductor must develop a plan to anticipate these issues and mitigate their effects.

1.4. Summary: Lessons Learned

Table 1 summarizes the issues and lessons learned from the data collection process, and proposes a resolution for each of the issues.

Table 1. Data Collection Process Issues, Lessons Learned, and Proposed Resolutions

Topic	Issue/Lesson Learned	Resolution
Data Synchronization	Different sampling rates were used in the collection of numerical and video data, resulting in inconsistencies between (non-synchronized) data.	Develop method to accurately and reliably synchronize the data collected, possibly using a video time-stamp to coordinate video data with timed numerical data.
Data Collection Error Mitigation	Process of collecting data from computer components on vehicles sometimes produces unusable and invalid data that are transferred to the independent evaluator.	Develop improved diagnostics for error-handling in raw data before data transfer to the evaluator, minimizing the occurrence of data errors due to sensor failure.
Data Duplication	Current flat-file format design by the FOT conductor is vulnerable to duplicated data.	Design an RDBMS that the FOT conductor and independent evaluator can use for organizing and analyzing data.
Data Integrity	Current flat-file format design requires additional software and debugging routines to test for errors, and is vulnerable to data corruption.	Develop a relational database that uses simpler, cleaner SQL scripts to test for data integrity that can be used to house data retrieved from the FOT vehicle.
Non-Identical Database Design	The design of the independent evaluator database is not the same as the FOT conductor database. If data discrepancies are discovered by the conductor after the original data transfer, data fix requires a resend and reload to the evaluator database of all affected data.	FOT conductor and independent evaluator should use identical RDBMS to streamline the data transfer process.
Overall Database Structure	Data shared with the independent evaluator are in non-conforming file formats that make data processing, testing, and analysis inefficient.	Independent evaluator should design a relational database to streamline data normalization and analysis, and provide the conductor with the exact file format and schema for data transfer.

Communication Process	Communication between the FOT conductor and independent evaluator was insufficient for the data collection and exchange process.	The independent evaluator must be engaged in all the discussions related to the data, database structure, and design.
Conservation of Resources	The accumulated effect of the data issues outlined above amounts to an overall lesson learned: existing system and infrastructure do not optimize the efficient use of time and effort by the team members.	Employing an RDBMS will streamline the data collection and analysis process to save resources for both the FOT conductor and independent evaluator.

1.5. Reasons to Use RDBMS

It is recommended that the independent evaluator designs and develops a relational database, and coordinates this design with the FOT conductor, using RDBMS software.

RDBMS software controls data storage, retrieval, deletion/modifications, security, and data integrity within a database. There are several advantages to using an RDBMS, as opposed to a flat-file database system:

- In a relational database, the data are transparent. There is no need to design and write software to read/write data. In contrast, flat-file systems require developing and debugging routines to read and write to and from flat files. Also, these routines must be revisited when either the file structure changes or data fixes occur.
- Data fixes in a relational database system can be applied to target the variables or rows that have problems. By contrast, in a flat-file system, data files are always read from top to bottom, even if you need access only to certain rows or lines in the middle of the file.
- Tables in a relational database can easily be linked to other tables or even other databases and can be searched and read from any location within a file.

Three of the lessons learned (data duplication, non-identical database design, and overall database structure) cite mismatched database design as a source of problems in processing and exchanging FOT data. The FOT conductor and independent evaluator should share consistent schema of the relational database. This will make data transfer much easier and allow sharing of Structured Query Language (SQL) scripts to fix errors. A shared relational database would minimize the need to resend data. All efforts should be made to avoid resending data unnecessarily.

One of the strengths of a relational database is its simplicity in organizing and querying data. Moreover, data are organized by many tables. A table stores data in rows and columns similar to the structure of an Excel spreadsheet. Each column stores one type of data (e.g., integer, strings, date, and time). Each column represents a data variable and each row represents a data record.

A relational database may be implemented using several relational database management systems from different database vendors who all follow the American National Standards Institute (ANSI) standard. The following are examples of this compliance:

- Each vendor provides several features that are available only in their database software in addition to supporting the ANSI standards.
- A query language is used to interact with any relational database in SQL, and every vendor supports the basic ANSI SQL standards, in addition to proprietary features.

- Each system has its own advantages and disadvantages, with final selection dependent on technical requirements and budget.

The original data files collected by the DAS should never be modified, but instead, should be used only to import data into relational data structures and then archived. Importing data from flat files into a relational database system does not make the data clean and queries faster. The data brought in for analysis must follow relational database system rules. Once data are organized and stored in an efficient manner, data become transparent and can be queried for checks, and fixes can be made quickly by clear and simple SQL routines, as opposed to writing complicated software routines to fix errors.

Section 2 proposes a step-by-step solution for a relational database design.

2. RELATIONAL DATABASE MANAGEMENT SYSTEM REQUIREMENTS FOR FOT DATABASE DESIGN AND MAINTENANCE

This section discusses a relational database management system (RDBMS) to be designed by the independent evaluator in coordination with the data collection needs of the FOT conductor. The RDBMS design should be agreed upon and used by both the FOT conductor and the independent evaluator.

The subsections are as follows:

- 2.1 describes the requirements and design for a relational database
- 2.2 discusses data exchange between the FOT conductor and the independent evaluator
- 2.3 describes how data are exported from the FOT conductor data files to the relational database files used by the independent evaluator
- 2.4 describes the components that can be used for data exchange and export

2.1. FOT Relational Database Requirements and Design

This subsection describes the steps in the process of identifying requirements and designing a relational database. It is recommended that Microsoft SQL Server software be used for relational database design in future field tests.

In a field test, there are often hundreds of variables, different data types, and data sampling at different rates resulting in millions or even billions of rows. Below are the basic steps to consider when designing a database to store and analyze field test data:

1. Identify the purpose and analysis needs of the FOT database, including the design of the user interface for the analysis and of the analysis reports.
2. Identify and categorize the list of variables to be collected and recorded in the FOT.
3. Design database tables that will hold the data and logical schema that will govern the overall database design and the relationships among the tables.
4. Identify the primary keys that will be used to uniquely identify data within each table, and the foreign keys that will establish relationships between tables.
5. Identify data integrity rules and relationship constraints.
6. Identify and plan for metadata storage.
7. Normalize the data and the database design.
8. Test and finalize the design.

Step 1: Identify the purpose and analysis needs of the FOT database

Before a database can be designed, the FOT conductor and the independent evaluator must decide what data will be stored and how they will be organized based on the needs of the evaluation. A database design must be planned at the early stages of the field test, once data channel lists and collection sampling rate details become available.

The FOT conductor and independent evaluator must understand the stated analysis goals and anticipate other likely database needs to ensure they conform to the database principles. They should have an understanding of the data types to be collected and address the limitations presented by data collection.

Step 2: Identify and categorize the list of variables to be collected and recorded in the Field Test

The FOT conductor is responsible for gathering the initial information regarding DAS variables and system data recording. All the data types should be identified and synchronization of different data types should be addressed. For example, the FOT conductor must determine if numeric and video data can be collected in a synchronized manner as raw data from equipment sensors or if they must be synchronized in a post-collection process.

Step 3: Design database tables to hold data fields and logical schema that will govern the overall database design and relationships among the tables

A database architect will utilize the data information and analysis needs to develop a logical database schema and table design. If any refinements need to be made to data design decisions made in steps 1 and 2, they should be asserted and corrected early in the database design process.

An example of database table design follows: demographic information of field test subjects should be stored in the database using the following approach. While data are tied to subjects in the database, demographic information is not and there is no need to house this information along with other data in the database. For this purpose, a Driver table is created including the following columns:

- DriverID
- Age
- Age Group

Each row in this table would represent a driver. When a row is added to this table, values for all the columns would have to be provided.

See Section C.1 in Appendix C for a diagram of the Database Table Structure designed by the independent evaluator for the CICAS-V pilot FOT.

Step 4: Identify the primary keys used to uniquely identify data within each table, and the foreign keys that establish relationships between tables

Tables in a relational database have special fields called keys. A primary key could be a single column or combination of columns that uniquely identify a row within a table. In the example described in Step 3, a Driver table will have DriverID as the primary key column that identifies a driver uniquely. Another table that would relate to the Driver table would include the DriverID as a foreign key, used as a means to connect to the Driver table.

An example of this could be a table that houses data collected from a radar device, as the Table named T_VOL_RADAR_FRONT shown in Appendix C. This table would typically have information such as:

- FileID, FrameNumber
- Radar Track
- Range (X_Range, Y_Range)
- Range Rate (X_Velocity, Y_Velocity)

In this example, the radar is capable of tracking 12 targets with a range limit. When the data stream is read and written to files by the DAS software, there is always a possibility that the data are written incorrectly or the unit conversion is not done accurately; for instance, if the radar reports information for track number 13. This is not possible because the maximum number of radar tracks is 12.

When the data files are imported into the FOT database, the constraints will ensure the data are within the range. Basic error checking, like checking the range of values, is a requirement and should be automated at the database level. Other quality checks, like checking for frozen data or missing data, can also be ensured by writing SQL routines that are transparent. These checks will ensure the following:

- Invalid data or erroneous data are caught at the very basic level and measures are taken to fix them.
- Sending data with errors to the independent evaluator and then resending the corrected data can be avoided.

Step 5: Identify data integrity rules and relationship constraints

The next step of the database construction process involves enforcing data integrity constraints. These constraints are simple to enforce at the database level when they reside inside the database. The alternative to this approach is to write separate routines to perform quality checks, which must be enforced all the time. When data are downloaded from the instrumented vehicles and imported into a constraints-enabled database, the first step is to capture errors along with the error cause. From this step, an appropriate action can be taken to make changes to the range of values and/or correct the DAS software. This type of constraint enforcing and checking for data quality must be in place from the very first stage of data collection and followed throughout the field test.

Step 6: Identify and plan for metadata storage

Metadata are summary information derived from collected data and used to understand what is stored in the database. In Appendix C, Section C.1 depicts the Data Structure by the independent evaluator. The tables labeled Summary Tables contain metadata, derived from the Time-series Data/Radar Tables. Due to the size of the FOT database, it is very important for metadata information to be generated from the data and stored within the same relational database. Metadata become very useful when checking on a batch of data to see if it passed quality checks, how many trips were invalid, trip duration, start/end data, and time of a trip.

Step 7: Normalize the data

Data are normalized by the process of organizing the data in the database to maximize the efficiency of how data are stored and used. In normalization, the data are tested to make sure that:

- There are no redundant data or duplicate columns within a database table.
- The data dependencies are logical to the purpose of the database.
- The relationships between tables work properly to support the purpose of the database.

The database designer should follow accepted guidelines for the process of normalizing data.

Step 8: Test and finalize the database design

When the design of the field test database structure is completed, the databases of the FOT conductor and the independent evaluator will have identical structures. SQL scripts will then allow data to be exchanged between them. Any changes to data within the finalized tables will not be possible at this stage, but additional tables can be created to add new variables. Any change to the database structure will need to follow protocols and must be reported to all parties because the dependencies based on the database may be affected.

Figure C-1 in Appendix C illustrates a diagram of the data structure used by the independent evaluator in the CICAS-V pilot field test.

2.2. Data Exchange

The FOT conductor will transfer data to the independent evaluator in batches. Depending on the type of field test, the evaluator may prefer to receive data from subjects after they have completed their full FOT period. Before any FOT data transfer can begin, it is important to have an efficient and well-tested method in place to exchange and import large quantities of data so as to identify and resolve any problems.

The administrator of the independent evaluation database will set up the database environment (e.g., server, database software, operating system) in a way that is appropriate for specific and individual applications that will access the FOT databases. Since the database environments at the independent evaluator and the FOT conductor may be different, it is recommended that there be a mutual understanding about the best way to exchange data from one environment to the other. Specifically, the independent evaluator and the FOT conductor must agree on how to export and import the FOT data. Microsoft SQL Server has been used as the relational database management system. SQL Server supports several file formats for exporting data. The exported data files in SQL Server can be in their native or text format as long as the export adheres to the database schema.

Section C.2 in Appendix C provides a flowchart depicting the process of data transfer from the FOT DAS to the independent evaluator.

2.3. Data Export

Data exporting is a process of extracting data from many tables in a database from an instance (database environment) of SQL Server into many user-defined files. For example, a flat file is created by copying rows from database table data.

When dealing with very large databases, it is best to organize and export data by subject and table type (e.g., Data, Data2). This will ensure that a particular exported data file will contain data from only a single subject as opposed to mixing data from different subjects. Since exporting data will not be a one-time occurrence, a set of export scripts must be developed and tested so that all files created are similar in structure and format.

SQL Server provides tools and utilities to export data. It is recommended that bcp, or bulk copy, utility be used since it is the fastest method. The use of this method ensures data integrity and also generates an optional format file that can be used by the evaluator's target database for importing data.

The following guidelines should be used for the SQL Server to optimize data export:

- Data are sorted by primary key in the exported file. For example, if Driver, Trip, and Time make up the primary key, then sort and export data by the primary key.

- Data are exported into several data files as opposed to a single large file. Using the earlier example, export data by Driver per table.
- Data file format is either in native or text format with row and column delimiters.

To illustrate these points, if there are ten tables in the database, exchanging data for two subjects would result in twenty export files, one per table per driver. The exported data files would have to follow a naming convention (FOTName-xx-yyy-zzz.dat), as shown in Table 2 so that export/import script generation and parsing could be automated.

For example, if Driver #1 data are being exported for tables Data and Data2, then the filenames would be as follows: FOTName-001-Data.dat and FOTName-001-Data2.dat.

Table 2. Example of Export Data File

File Part	Description
FOTName	System Name
Xx	FOT Platform
Yyy	Driver #
Zzz	Table name

Appendix C provides more details about the final FOT database table structure, schema, and implementation.

2.4. Data Exchange Medium

The most common medium used for copying and exchanging any type of files is external USB drives. Today, every system comes with one or more USB ports, eliminating the need for an add-on card to support such media. In theory, the maximum transfer rate for USB 2.0 is 60 MB/s; but in practice, the rate does not exceed 30 MB/s. This transfer rate will work for most regular user operations and data transfer. When USB drives are used to exchange large quantities of data, the copying speed is not fast enough.

Typically, only a high-speed data transfer interface is used on servers and workstations. There are several interfaces that could be used to connect storage to get high bandwidth, but currently the interface most commonly used and deployed in data centers and workstations is SAS (Serial-Attached SCSI). This interface supports high-speed enterprise drives (SAS-to-SAS 7.2K/15K RPM), as well as high-capacity enterprise/desktop drives (SAS-to-SATA 7.2K RPM). Newer generation high-capacity drives are available up to 1 TB, and they are all supported using this interface. The maximum bandwidth of a SAS/SATA interface is 300 MB/s, but the only drives available that support this rate are Intel SSDs (Solid-State Drives); however, these drives are prohibitively expensive for field test data exchange.

The SAS/SATA interface typically transfers at the speed supported by the physical hard drive. For example, if a high-capacity drive's sustained rate is 80 MB/s, a similar speed could be achieved while copying files from a single drive, but this would under-utilize the bandwidth supported (300 MB/s). SAS and SATA drives that support all drive bandwidth capacities can be used at the enterprise level as well as desktop storage level, eliminating the need to have separate enclosures for different drive technologies and capacities.

2.4.1. Recommended Hardware Configuration for the SAS Interface

The external SAS interface (or port) for SAS enclosures are referred to as mini-SAS. An add-on card is required to have the mini-SAS interface on servers and workstations. Typically, four drives can be connected per mini-SAS port, providing 4×80 MB/s transfer rate. This allows a transfer rate of 320 MB/s, using above example, and would utilize the SAS bandwidth to the maximum.

The enclosure shown in Figure 1 supports drives on trays (picture on the right above shows the tray with the drive installed) that can be installed and removed easily. This also allows one to increase the storage capacity by just installing a new drive in the same enclosure to increase storage capacity. A regular external USB enclosure comes with only one drive restricting the storage capacity to the drive capacity. If more storage is needed, typically a new external drive with higher capacity will have to be bought. With the proposed enclosures, higher capacity drives can be added to the enclosure at any time to increase storage capacity.

Figure 1. Desktop SAS Enclosure

Based on the recommended hardware and experiences with other field tests, six SATA drives, each up to 2 TB in capacity, should be purchased for data exchange. Two drives should be at each location so that data can be shipped without waiting for drives to be sent back immediately. The two remaining drives are spares which should be kept by the independent evaluator in case a drive fails so the data exchange tasks are not affected.

The data exchange method and medium discussed above were introduced by the Volpe Center as the independent evaluator in the Integrated Vehicle-Based Safety System (IVBSS) field tests, and have been very successful. Appendix D highlights the experience of the Volpe Center as the independent evaluator of many FOT projects.

APPENDIX A. PILOT FOT DATA RECEIVED BY THE EVALUATOR

Below are examples of data received in flat-file format from the FOT conductor. The independent evaluator received sample data in a flat-file (.txt) along with a video file.

Below is how the variables are listed in the text-file.

```
Column   1: System.Frame_Number (None)
Column   2: OBE-Communication.ABS_Active (None)
Column   3: OBE-Communication.Brakes_Active (None)
Column   4: OBE-Communication.Stability_Control_Active (None)
Column   5: OBE-Communication.Driver_Intended_Brake_Level (None)
Column   6: OBE-Communication.Brake_Pedal_Position (None)
Column   7: OBE-Communication.Yaw_Rate (degrees/second)
Column   8: OBE-Communication.Outside_Air_Temperature (celcius)
Column   9: OBE-Communication.Vehicle_Speed (kilometers/hour)
Column  10: OBE-Communication.Lateral_Acceleration (meters/second/second)
Column  11: OBE-Communication.Panic_Brake_Active (None)
Column  12: OBE-Communication.PreCharge_Status (None)
Column  13: OBE-Communication.Headlight_Status (None)
Column  14: OBE-Communication.Wiper_Status (None)
Column  15: OBE-Communication.Longitudinal_Acceleration (meters/second/second)
Column  16: OBE-Communication.Vertical_Acceleration (meters/second/second)
Column  17: OBE-Communication.Left_Turn_Signal (None)
Column  18: OBE-Communication.Right_Turn_Signal (None)
Column  19: OBE-Communication.Accelerator_Pedal_Position (percent)
Column  20: OBE-Communication.Cruise_Set_Speed (kilometers/hour)
Column  21: OBE-Communication.Seatbelt_Status (None)
Column  22: OBE-Communication.Horn_Status (None)
Column  23: OBE-Communication.ACC_Status (None)
Column  24: OBE-Communication.ACC_Set_Speed (kilometers/hour)
Column  25: OBE-Communication.Intersection_ID (ID)
Column  26: OBE-Communication.Present_Approach (approach)
Column  27: OBE-Communication.Present_Lane (lane)
Column  28: OBE-Communication.Time_To_Intersection (milliseconds)
Column  29: OBE-Communication.Algorithm_Status (None)
Column  30: OBE-Communication.Current_Signal_Phase (None)
Column  31: OBE-Communication.Time_To_Next_Phase (milliseconds)
Column  32: OBE-Communication.Vehicle_Latitude (degrees)
Column  33: OBE-Communication.Vehicle_Longitude (degrees)
Column  34: OBE-Communication.Vehicle_Heading (degrees)
Column  35: OBE-Communication.Vehicle_Altitude (meters)
Column  36: OBE-Communication.Local_GPS_TOW (milliseconds)
Column  37: OBE-Communication.Local_GPS_Speed (meters/second)
Column  38: OBE-Communication.Local_GPS_Week_Number (gps_week)
Column  39: OBE-Communication.Distance_to_Stop_Bar (meters)
Column  40: HEL_CAN_Port.Video_Frame (None)
Column  41: Crossbow.Yaw_Rate (degrees/seconds)
Column  42: Crossbow.X_Accel (g)
Column  43: Crossbow.Y_Accel (g)
Column  44: Crossbow.Z_Accel (g)
Column  45: GPS.Weeks (weeks)
Column  46: GPS.Time (seconds)
```

```
Column  47: GPS.Status (None)
Column  48: GPS.Latitude (radians)
Column  49: GPS.Longitude (radians)
Column  50: GPS.Altitude (meters (HAE))
Column  51: GPS.Speed (meters/seconds)
Column  52: Front_Radar_SMS.Number_Of_Objects (objects)
Column  53: Front_Radar_SMS.Object_ID_T0 (ID)
Column  54: Front_Radar_SMS.X_Range_T0 (meters)
Column  55: Front_Radar_SMS.Y_Range_T0 (meters)
Column  56: Front_Radar_SMS.X_Velocity_T0 (meters/second)
Column  57: Front_Radar_SMS.Y_Velocity_T0 (meters/second)
Column  58: Front_Radar_SMS.Object_ID_T1 (ID)
Column  59: Front_Radar_SMS.X_Range_T1 (meters)
Column  60: Front_Radar_SMS.Y_Range_T1 (meters)
Column  61: Front_Radar_SMS.X_Velocity_T1 (meters/second)
Column  62: Front_Radar_SMS.Y_Velocity_T1 (meters/second)
Column  63: Front_Radar_SMS.Object_ID_T2 (ID)
Column  64: Front_Radar_SMS.X_Range_T2 (meters)
Column  65: Front_Radar_SMS.Y_Range_T2 (meters)
Column  66: Front_Radar_SMS.X_Velocity_T2 (meters/second)
Column  67: Front_Radar_SMS.Y_Velocity_T2 (meters/second)
Column  68: Front_Radar_SMS.Object_ID_T3 (ID)
Column  69: Front_Radar_SMS.X_Range_T3 (meters)
Column  70: Front_Radar_SMS.Y_Range_T3 (meters)
Column  71: Front_Radar_SMS.X_Velocity_T3 (meters/second)
Column  72: Front_Radar_SMS.Y_Velocity_T3 (meters/second)
Column  73: Front_Radar_SMS.Object_ID_T4 (ID)
Column  74: Front_Radar_SMS.X_Range_T4 (meters)
Column  75: Front_Radar_SMS.Y_Range_T4 (meters)
Column  76: Front_Radar_SMS.X_Velocity_T4 (meters/second)
Column  77: Front_Radar_SMS.Y_Velocity_T4 (meters/second)
Column  78: Front_Radar_SMS.Object_ID_T5 (ID)
Column  79: Front_Radar_SMS.X_Range_T5 (meters)
Column  80: Front_Radar_SMS.Y_Range_T5 (meters)
Column  81: Front_Radar_SMS.X_Velocity_T5 (meters/second)
Column  82: Front_Radar_SMS.Y_Velocity_T5 (meters/second)
Column  83: Front_Radar_SMS.Object_ID_T6 (ID)
Column  84: Front_Radar_SMS.X_Range_T6 (meters)
Column  85: Front_Radar_SMS.Y_Range_T6 (meters)
Column  86: Front_Radar_SMS.X_Velocity_T6 (meters/second)
Column  87: Front_Radar_SMS.Y_Velocity_T6 (meters/second)
Column  88: Front_Radar_SMS.Object_ID_T7 (ID)
Column  89: Front_Radar_SMS.X_Range_T7 (meters)
Column  90: Front_Radar_SMS.Y_Range_T7 (meters)
Column  91: Front_Radar_SMS.X_Velocity_T7 (meters/second)
Column  92: Front_Radar_SMS.Y_Velocity_T7 (meters/second)
Column  93: Rear_Radar_SMS.Object_ID_T0 (ID)
Column  94: Rear_Radar_SMS.X_Range_T0 (meters)
Column  95: Rear_Radar_SMS.Y_Range_T0 (meters)
Column  96: Rear_Radar_SMS.X_Velocity_T0 (meters/second)
Column  97: Rear_Radar_SMS.Y_Velocity_T0 (meters/second)
Column  98: Rear_Radar_SMS.Object_ID_T1 (ID)
Column  99: Rear_Radar_SMS.X_Range_T1 (meters)
Column 100: Rear_Radar_SMS.Y_Range_T1 (meters)
Column 101: Rear_Radar_SMS.X_Velocity_T1 (meters/second)
```

```
Column 102: Rear_Radar_SMS.Y_Velocity_T1 (meters/second)
Column 103: Rear_Radar_SMS.Object_ID_T2 (ID)
Column 104: Rear_Radar_SMS.X_Range_T2 (meters)
Column 105: Rear_Radar_SMS.Y_Range_T2 (meters)
Column 106: Rear_Radar_SMS.X_Velocity_T2 (meters/second)
Column 107: Rear_Radar_SMS.Y_Velocity_T2 (meters/second)
Column 108: Rear_Radar_SMS.Object_ID_T3 (ID)
Column 109: Rear_Radar_SMS.X_Range_T3 (meters)
Column 110: Rear_Radar_SMS.Y_Range_T3 (meters)
Column 111: Rear_Radar_SMS.X_Velocity_T3 (meters/second)
Column 112: Rear_Radar_SMS.Y_Velocity_T3 (meters/second)
Column 113: Rear_Radar_SMS.Object_ID_T4 (ID)
Column 114: Rear_Radar_SMS.X_Range_T4 (meters)
Column 115: Rear_Radar_SMS.Y_Range_T4 (meters)
Column 116: Rear_Radar_SMS.X_Velocity_T4 (meters/second)
Column 117: Rear_Radar_SMS.Y_Velocity_T4 (meters/second)
Column 118: Rear_Radar_SMS.Object_ID_T5 (ID)
Column 119: Rear_Radar_SMS.X_Range_T5 (meters)
Column 120: Rear_Radar_SMS.Y_Range_T5 (meters)
Column 121: Rear_Radar_SMS.X_Velocity_T5 (meters/second)
Column 122: Rear_Radar_SMS.Y_Velocity_T5 (meters/second)
Column 123: Rear_Radar_SMS.Object_ID_T6 (ID)
Column 124: Rear_Radar_SMS.X_Range_T6 (meters)
Column 125: Rear_Radar_SMS.Y_Range_T6 (meters)
Column 126: Rear_Radar_SMS.X_Velocity_T6 (meters/second)
Column 127: Rear_Radar_SMS.Y_Velocity_T6 (meters/second)
Column 128: Rear_Radar_SMS.Object_ID_T7 (ID)
Column 129: Rear_Radar_SMS.X_Range_T7 (meters)
Column 130: Rear_Radar_SMS.Y_Range_T7 (meters)
Column 131: Rear_Radar_SMS.X_Velocity_T7 (meters/second)
Column 132: Rear_Radar_SMS.Y_Velocity_T7 (meters/second)

End Data Header
```

APPENDIX B. DATA CORRECTION PROCESS

B.1. CICAS-V Data Processing Steps

- The independent evaluator created a new relational database in SQL Server to accommodate and review the data sample.
- Using a list of variables (from the text-file), the independent evaluator created a single "table" manually defining the variables and their data types by guessing based on the units-of-measurement.
- The independent evaluator identified and defined the primary key.
- The independent evaluator imported the data into the table with a primary key in place.

B.2. Data Issues

- Two data issues are illustrated by the data in Table B-1 below. (Line numbers below correspond to the numbers in the boxes to the left of the table):

- Duplicate frame numbers were found in the dataset, shown by the first oval drawn around the repeated Frames 161 and 162.

- For Frames 161 through 169, when the variable "NumberOfObjects" was 0 (indicated by the second oval) for "front radar," values were seen for "radar" variables.

Table B-1. Duplicate Data

Frame	NumOfObjects	T0 ObjectID	XRange	YRange	XVelocity	YVelocity	T1 ObjectID	XRange	YRange	XVelocity	YVelocity
155	3	26	0.58	131.3	51.1	-76.9	26	0.19	131.74	51.1	-77
156	2	26	0.13	131.39	-51.2	-76.9	27	2.4	131.68	-51.6	-76.9
157	1	27	1.82	131.71	50.8	-77	27	2.4	131.68	-51.6	-76.9
158	1	27	1.22	131.78	50.8	-77	27	2.4	131.68	-51.6	-76.9
159	1	27	0.67	-130.24	-51.5	-77	27	2.4	131.68	-51.6	-76.9
160	1	27	0.32	-130.14	-51.5	-77	27	2.4	131.68	-51.6	-76.9
161	0	27	0.32	-130.14	-51.5	-77	27	2.4	131.68	-51.6	-76.9
162	0	27	0.32	-130.14	-51.5	-77	27	2.4	131.68	-51.6	-76.9
161	0	27	0.32	-130.14	-51.5	-77	27	2.4	131.68	-51.6	-76.9
162	0	27	0.32	-130.14	-51.5	-77	27	2.4	131.68	-51.6	-76.9
163	0	27	0.32	-130.14	-51.5	-77	27	2.4	131.68	-51.6	-76.9
164	0	27	0.32	-130.14	-51.5	-77	27	2.4	131.68	-51.6	-76.9
165	0	27	0.32	-130.14	-51.5	-77	27	2.4	131.68	-51.6	-76.9

166	0	27	0.32	-130.14	-51.5	-77	2.4	131.68	-51.6	-76.9
167	0	27	0.32	-130.14	-51.5	-77	2.4	131.68	-51.6	-76.9
168	0	27	0.32	-130.14	-51.5	-77	2.4	131.68	-51.6	-76.9
169	0	27	0.32	-130.14	-51.5	-77	2.4	131.68	-51.6	-76.9

- A third data issue is illustrated by the data shown in Table B-2, below. The data table displays variables for two available targets, T0 and T1, designed to hold the same type of radar data variables (ObjectID, XRange, YRange, XVelocity and YVelocity) according to the number of objects targeted. The data shown below are invalid because, where the variable "NumberOfObjects" reports 1 (located in the second column), "radar data" were available for more than 1 target (T0 and T1). A valid dataset would show only one radar target if NumOfObjects equals 1.

Table B-2. Invalid Data Rows

Frame	NumOfObjects	T0					T1				
		ObjectID	XRange	YRange	XVelcity	YVelcity	ObjectID	XRange	YRange	XVelcity	YVelcity
143	2	26	0.13	131.3	51.3	-76.7	26	3.01	-130.94	-51.4	-76.6
144	1	26	2.53	-130.91	-51.4	-76.6	26	3.01	-130.94	-51.4	-76.6
145	2	26	2.05	131.3	-51.4	-76.6	26	4.35	130.94	-51.5	-76.8
146	2	26	1.73	131.33	-51.4	-76.6	26	4.13	130.98	-51.5	-76.8
147	2	26	1.31	-130.78	51.1	-76.7	26	3.62	131.04	-51.5	-76.8
148	2	26	0.9	131.39	51.1	-76.7	26	3.1	-131.04	-51.4	-76.8
149	2	26	0.51	-130.78	-51.2	-76.7	26	2.78	-130.98	-51.4	-76.8
150	1	26	2.3	131.23	51	-76.9	26	2.46	-130.94	51	-76.9
151	1	26	1.98	131.26	51	-76.9	26	2.46	-130.94	51	-76.9
152	2	26	1.5	-130.75	-51.3	-76.9	26	1.92	131.1	51	-76.9
153	2	26	0.99	-130.62	-51.3	-76.9	26	1.47	131.17	51	-76.9

3

APPENDIX C. FINAL FOT DATABASE TABLE STRUCTURE, SCHEMA, AND IMPLEMENTATION

C.1. Independent Evaluation Data Structure

After reviewing the text-file, the independent evaluator designed a relational database to accommodate the data efficiently, adding few tables and variables. The independent evaluator identified and defined the primary key candidates based on their understanding of the data.

The diagram on the following page in Figure C-1 depicts the FOT database structure designed by the independent evaluator to store the pilot FOT vehicle DAS data. The diagram displays for each table, the field names in the first column, with the key fields shown first and in bold, and each field's data type in the second column, and the table relationships indicated by the arrows between tables:

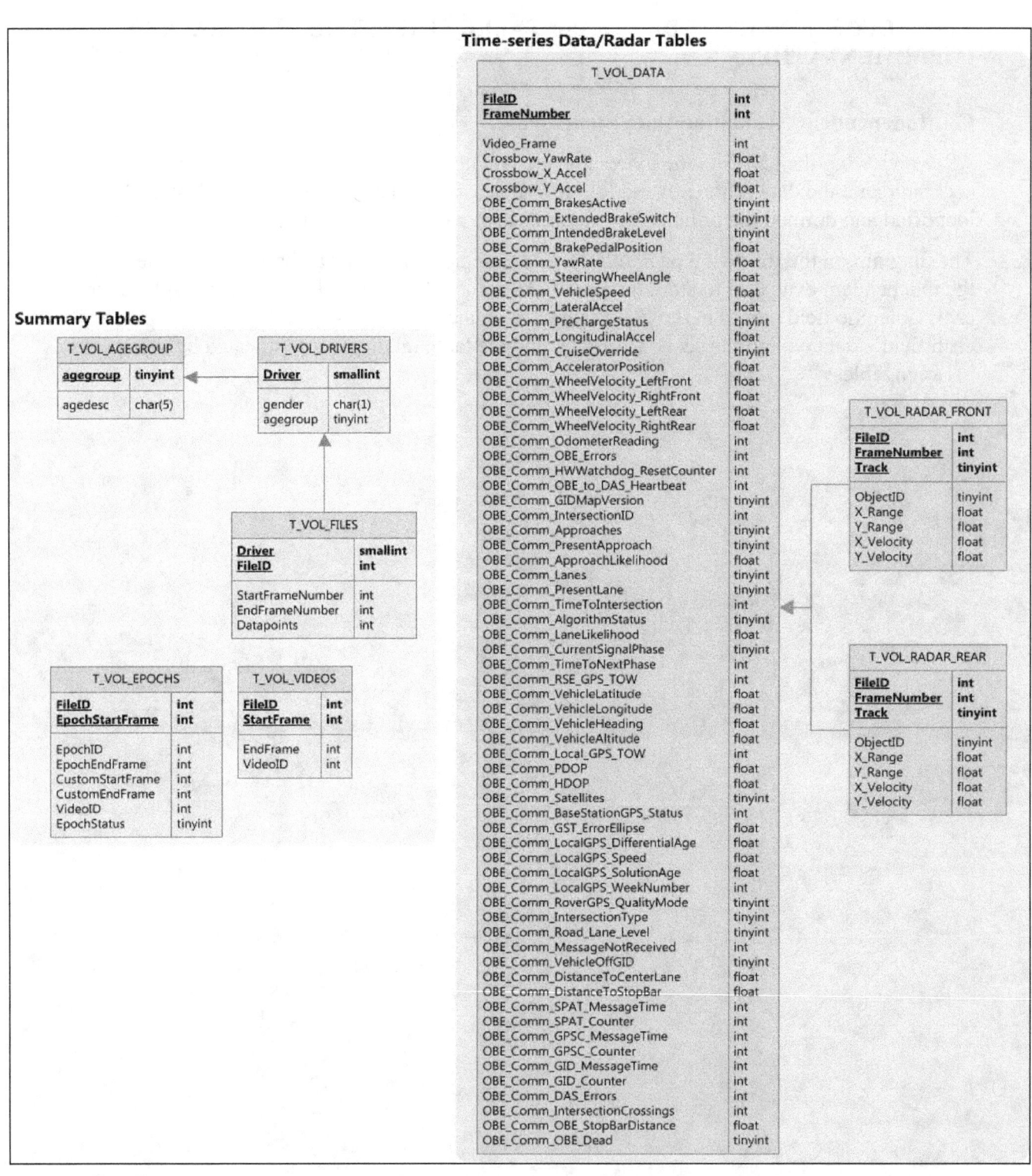

Figure C-1. Independent Evaluator Pilot FOT Database Table Structure

C.2. Data Exchange Flowchart

The task flowchart in Figure C-2 shows the process by which data are transferred from the FOT DAS to the independent evaluator for analysis, using a relational database as the means of data exchange.

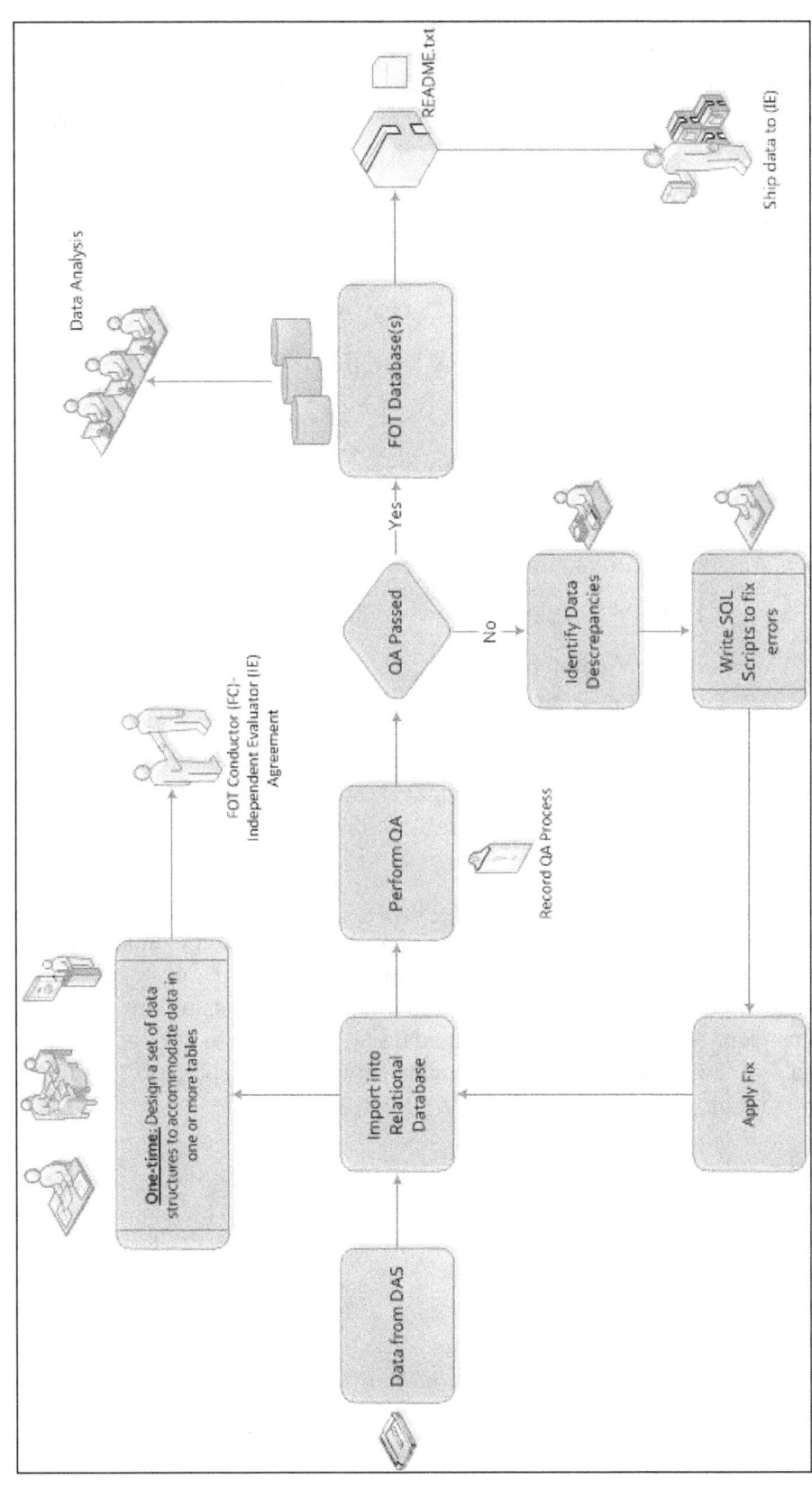

Figure C-2. Data Exchange Process from the FOT Vehicle DAS to Independent Evaluator

APPENDIX D. THE VOLPE CENTER'S FOT EXPERIENCE

As the independent evaluator, the Volpe Center has experience working with six different field tests and various FOT conductors. Below is the list of FOTs and supporting data collection tasks:

- Intelligent Cruise Control (ICC)
- Automotive Collision Avoidance System (ACAS)
- Roadway Departure Crash Warning System (RDCW)
- Drowsy Driver Warning System (DDWS)
- Integrated-Vehicle Based Safety Systems (IVBSS)
    - IVBSS Pilot
    - IVBSS Heavy Truck (HT) FOT
    - IVBSS Light Vehicle (LV) FOT
- Cooperative Collision Avoidance System for Violations (CICAS-V)
    - CICAS-V (Signal)
    - CICAS-V (Stop Sign)
    - CICAS-V (Pilot)

As the independent evaluator on each of the above FOT initiatives, the Volpe Center designed, developed, and implemented tools and databases to successfully conduct an independent evaluation. The experience gained through each progressive FOT was leveraged on subsequent FOTs.

Beyond database design, layout, implementation, and population, there are a number of other database-related development requirements for an independent evaluation. Table D-1 compares Prior FOTs of which the Volpe Center was the evaluator to database tools and attributes. Table D-2 compares CICAS-V data analysis tasks to database tools and attributes. Engineering algorithms were developed to process the data for analysis. As part of this analysis, a Multimedia Data Analysis Tool (MDAT) was also built to assist the independent evaluation team to analyzing videos.

Table D-1. Comparison of Prior Field Tests

FOT Initiative	Algorithms	Processing Tools	MDAT	Relational DB	Flat File	Video
ICC	New	N/A	N/A	Yes	No	.avi
ACAS	New	New	New	Yes	No	2xBinary New
RDCW	New/Partial	New	New	Yes	No	2xBinary New
DDWS	New	New	New	IE Ported to DB	Yes	Mpeg-2
IVBSS Pilot	N/A	New	New	Yes	No	2xRDCW Binary
IVBSS HT	New	New	New	Yes	No	5xBinary/Mpeg-4
IVBSS LV	Reused/ New	Reused	Reused	Yes	No	5xBinary/Mpeg-4

As you can see from the above table, the Volpe Center on the IVBSS FOT initiative planned and designed the tools and algorithms so that it can be reused to the extent possible to save time and effort. The IVBSS LV uses most of the algorithms used for IVBSS HT; but due to the specifications being different for LV, few algorithms were developed. Such planning should be exercised by the FOT conductors as well for all FOT initiatives. Special workstations were also needed to process video data specifically on all the above FOT initiatives.

Table D-2. Comparison of CICAS-V Field Tests

FOT Initiative	Algorithms	Processing Tools	MDAT	Relational DB	Flat File	Video
CICAS-V (Signal)[1]	New	New	New	IE Ported to DB	Yes	Mpeg-4
CICAS-V (Stop Sign)[2]	New	New	New	IE Ported to DB	Yes	Mpeg-4
CICAS-V Pilot[3]	N/A	New	New	IE Ported to DB	Yes	Mpeg-4

[1] CICAS-V Pilot – Data were collected on a CICAS-V equipped vehicle as it was driven over a prescribed route by tests subjects through several CICAS-V equipped signalized and stop sign controlled intersections.

[2] CICAS-V Stop Sign – Data collected from infrastructure installed at stop sign controlled intersections with historical crash data indicating violation related crashes.

[3] CICAS-V Signal– Data collected from infrastructure installed at signal controlled intersections with historical crash data indicating violation related crashes.

DOT HS 811 498
July 2011